SCIENTIFIC AMERICAN EDUCATIONAL PUBLISHING

EXPLORE PHYSICS

10 FUN PHYSICAL SCIENCE EXPERIMENTS

BRING SCIENCE HOME

Published in 2025 by Scientific American Educational Publishing in association with
The Rosen Publishing Group.

Contains material from Scientific American, a division of Springer Nature America, Inc., reprinted by permission as well as original material from The Rosen Publishing Group.

First Edition

Editor: Kristen Rajczak Nelson
Designer: Rachel Rising

Activity on p. 5 by Science Buddies/Sabine De Brabandere (September 6, 2018); p. 10 by Science Buddies/Ben Finio (August 23, 2018): p. 14 by Science Buddies/Megan Arnett (October 19, 2017); p. 21 by Science Buddies/Ben Finio (August 3, 2017); p. 25 by Science Buddies/Megan Arnett (February 9, 2017): p. 31 by Science Buddies (April 11, 2013); p. 37 by Science Buddies (January 24, 2013); p. 43 by Science Buddies (November 26, 2015); p. 49 by Science Buddies (November 26, 2015); p. 55 by Science Buddies/Ben Finio (December 26, 2019).

All illustrations by Continuum Content Solutions

Photo Credits: pp. 3-5, 8, 14, 19, 37, 41, 43, 47, 49, 53 cve iv/Shutterstock.com; pp. 5, 10, 14, 21, 25, 31, 37, 43, 49, 55 Anna Frajtova/Shutterstock.com.

Library of Congress Cataloging-in-Publication Data
Names: Scientific American, inc., editor.
Title: Explore physics : 10 fun physical science experiments / Scientific American editors.
Description: Buffalo, New York : Scientific American Educational Publishing, an imprint of Rosen Publishing, [2025] | Series: Bring science home | Includes index.
Identifiers: LCCN 2024005874 (print) | LCCN 2024005875 (ebook) | ISBN 9781725350588 (library binding) | ISBN 9781725350571 (paperback) | ISBN 9781725350595 (ebook)
Subjects: LCSH: Physics--Experiments--Juvenile literature. | Physics--Juvenile literature. | Physical sciences--Experiments--Juvenile literature. | Physical sciences--Juvenile literature. | Science projects--Juvenile literature. | Science--Experiments--Juvenile literature.
Classification: LCC QC25 .E97 2025 (print) | LCC QC25 (ebook) | DDC 530.078--dc23/eng/20240229
LC record available at https://lccn.loc.gov/2024005874
LC ebook record available at https://lccn.loc.gov/2024005875

Manufactured in the United States of America

CPSIA Compliance Information: Batch #CSSA25. For further information contact Rosen Publishing, New York, New York at 1-800-237-9932.

CONTENTS

THESE ACTIVITIES INCLUDE SCIENCE FAIR PROJECT IDEAS.

INTRODUCTION

Finding physics in your everyday life isn't as hard to do as you might think! In these experiments, you'll see if air can plug a bottle, figure out how to balance anything with a circus-like twist, and see if you can build your own cargo-carrying vessel. You'll see how physics is at the center of all of it!

Projects marked with ⚛ include a section called Science Fair Project Ideas. These ideas can help you develop your own original science fair project. Science fair judges tend to reward creative

Go with the Flow

GLUG-GLUG: USE PHYSICS—AND SOME SCIENTIFIC TESTING—TO FIND A BETTER WAY TO POUR LIQUIDS.

Do you ever wonder why we perform tasks in a particular way? Very often we learn by copying someone and never stop to question if there is a more efficient or easier way to do something. This activity is about the tricky task of pouring from a full container. Could there be an easier way to do it?

PROJECT TIME

20 minutes

KEY CONCEPTS

Physics
Fluids
Flow
Process

BACKGROUND

Almost everything around us is either a solid (such as a wooden block), a liquid (like water you drink), or a gas (as in the air you breathe). These are all made up of tiny particles, but the strength by which they connect to one another differs. In solids, these particles hold on to one another tightly. For liquids and gases, these particles can flow over or alongside one another. That is why liquids and gases are also called fluids: because they can flow. This flow can be smooth, chaotic, or anything in between. A smooth flow is called laminar, and its opposite is called a turbulent.

When you pour fluid out of a container, you remove particles from that container leaving space behind. One of two things can happen as a result: The container can collapse and become smaller or another fluid can rush in and fill the emptiness. This happens because the tiny particles that make up a fluid press against everything around them. As long as the particles inside the container press about as hard as the particles on the outside, the container will hold its form. If fluid flows out, there are suddenly many fewer particles pressing on the inside of the container while the pressure on the outside remains constant. As a result, the container might collapse.

In other circumstances, another fluid will rush in, fill the space, and help press against the inner walls so the inside and outside pressures equalize again. This activity explores the different ways a fluid can flow in while you pour water—and how that can affect the flow.

MATERIALS

- Empty rectangular soup or milk box (such as shelf-stable soy milk) with a spout on top
- Water
- Marker
- Glass or cup to pour into
- Paper towels or rag (to dry up spills)
- A workspace that can tolerate water spills
- Paper and pen (optional)

PREPARATION

- The box has six sides: one on the top with the spout, one on the bottom and four side panels that connect the top and the bottom. Write the letter A on the side panel farthest from the spout. Go around the box, lettering the next side panel B, then C and, finally, D.

PROCEDURE

- Fill the box with water. The water level should be about 1/2 inch (1.3 cm) from the top.
- Pour some of the box's water rather quickly into a glass in the way you usually do. Pay attention. *Which side panel do you grip while you pour? Is the flow fast or slow? Does the water flow fluently? Do you think there are other ways to pour? If so, would those work better, as well or worse?*
- If you poured quite slowly, repeat the previous steps but pour faster and see how this changes how the water flows.
- Empty the glass and add more water to the box so it is filled to the same level as before.
- Pour some water rather quickly into the glass, but this time hold the box by a different side panel. *Is this easier or harder compared with the way you usually do it? How does the water flow when you do it this way?*
- Repeat two more times, each time holding the box by a different side panel. Rank the ways you poured water from most to least preferred and from a laminar (or fluent) water flow to a turbulent (or chaotic) one. If you have paper and a pen, write down your ranking.
- Repeat the whole activity two more times. Pay attention to how far you tilt the box before water pours out and whether or not the spout is fully covered with water. *Do you get the same ranking each time?*
- *Can you explain why the water flows differently depending on which side you hold the box to pour?*

EXTRA

Do you think your findings are still true if the box is half full or nearly empty? Why? Try it and see if your prediction was correct!

EXTRA

Test whether your findings are the same if you pour really slowly.

EXTRA

Repeat the activity after making a hole in the top panel of the box. Keep using the spout to pour. *How does this change your rankings? Did it make pouring easier? Why?*

SCIENCE FAIR IDEA

Test out different water bottles. *Do some show turbulent flow when water is poured quickly? Some water bottles have a small hole near the spout in the lid—why would this be? Why does liquid not flow out from this small hole?*

OBSERVATIONS AND RESULTS

Did the water flow the smoothest when you held the box on side panel C? Did it flow most turbulently when your hand covered part of side panel A? Maybe you even developed a new preference. Preferences are personal choices and can differ from person to person. Laws of physics, by contrast, determine how the water flows.

You might think the box was empty when you poured all the water out, but it was not. Something else filled the box—something you cannot see: air. If liquid is poured out, another fluid such as air will take its place or the container will collapse. Unless you take very special measures, the empty space created by pouring out water does not stay empty for long! In your case, the container probably held its form and air rushed in.

Sometimes air can flow in, and water can pour out simultaneously. The two flow alongside each other, resulting in fluent or laminar flows. You probably saw this when you poured while gripping side panel C. Holding the box by this side allows you to tilt the box quite far before the water level reaches the spout. This makes it easier to create a nice stream of water without filling the spout opening entirely, leaving space for air to flow in while water pours out.

When you poured while holding side panel A, a much smaller tilt of the box made the water pour out but it almost instantly covered the spout completely, leaving no room for air to flow in. As water pours out and there is nothing to replace it, a pressure imbalance builds up. If this imbalance is great enough, the water flow can stop for a moment to suck in air, which results in water glugging out. Each glug is accompanied by a break for air to flow. This is an example of a chaotic or turbulent water flow, which is much harder to control.

CLEANUP

Dry off your workspace, clean the glass, and recycle the box if you can.

Can You Catch a Bubble?

LEARN HOW SURFACES MIGHT BE DIFFERENT THAN THEY SEEM WITH THIS PHYSICS ACTIVITY THAT REALLY POPS!

Have you ever tried to catch a bubble without popping it? It's hard! What materials can you use to successfully catch a bubble? Do some materials work better than others? Try this project to find out.

PROJECT TIME

20 to 30 minutes

KEY CONCEPTS

Physics
Surfaces
Surface tension
Hydrophilic
Hydrophobic

BACKGROUND

Bubbles are fun and beautiful—but also fragile! A bubble is made from a thin film of soapy water with air inside. Many different things, such as contact with a solid surface, can cause the film to break, popping the bubble. But it can even pop without touching anything because the water in it gradually evaporates, making the film weaker. Sometimes, however, you might notice a bubble can land without popping.

Whether a bubble pops when it comes in contact with a solid surface depends on many different factors, including the surface properties of the material. You probably know different materials have different properties, some of which you can see or feel (such as color or density). Materials, however, have other characteristics that are harder to observe directly. Surfaces can be hydrophobic (repel water) or hydrophilic (attract water). You can observe this by dropping water onto a surface and seeing whether it forms big beads (hydrophobic) or spreads out in thin sheets (hydrophilic). Whether a material is hydrophobic or hydrophilic depends strongly on its surface roughness. Some materials, such as sandpaper, have macroscopic surface features, meaning you can feel the bumps and see them with your naked eye. Other materials, however, have microscopic ones. Even if the material looks and feels smooth to you, it might have very tiny bumps or pores. These can actually help the material repel water because the latter is held together by surface tension, and thereby unable to penetrate into the tiny gaps in the material. Other materials such as paper or sponges have larger gaps that help them absorb water.

All of these properties, along with other factors such as how fast the bubble is moving and whether the surface is wet, can affect whether a bubble is more likely to pop when it lands on a particular surface. Try this project to find out what works best to catch a bubble!

MATERIALS

- Bubble solution (You can buy some or make your own at home with distilled water, dish soap, and glycerin or corn syrup. See the "Preparation" section if you'd like to do this.)
- Small bowl or container to hold your bubble solution if you make your own
- Bubble wand
- Tap water
- Paper towels
- Tape (optional)

- Different materials to test such as aluminum foil; wax paper; printer paper; plastic wrap; wood; plastic bags or food storage containers; metal pots and pans; tables; countertops; etcetera (You can try any material you find around your house that can get soapy water on it! **Note:** You might not want to do this activity with certain materials such as wood or furniture because water could damage them. Check with an adult if you aren't sure.)

PREPARATION

- To make your own bubble solution, mix 1 cup of water (237 ml) with 2 tablespoons (30 ml) of dish soap and 1 tablespoon (15 ml) of glycerin or corn syrup.
- Gather the different materials you want to test. Lay them out on a flat surface, preferably indoors. (It will be difficult to do this project outdoors on a windy day.)

PROCEDURE

- Dip your bubble wand in your bubble solution and gently blow some bubbles toward your first sheet of material. Try to get them to land on it and not go past it. This may take some practice, so try it a few times. *What happens when the bubbles hit the surface?*
- Try blowing bubbles onto each of your other materials. *Do the bubbles land on any of them without popping?*
- Wet a paper towel; use it to wipe a thin layer of water onto each surface.
- Try blowing bubbles onto each surface again. *What happens now?*
- Dip a new paper towel into your bubble solution; wipe a thin layer of it onto each test surface.

- Blow bubbles onto each surface again. *Do the bubbles stick to any surfaces where they previously popped? Do they last longer after they land?*

EXTRA

Does the orientation of the surface matter? Try using some tape to put your materials on a wall or even on the ceiling (checking for permission first before taping onto any surface). *Can you still get bubbles to land on them?*

OBSERVATIONS AND RESULTS

When all your materials were dry, you probably found smooth, waterproof surfaces did the best job catching bubbles. Wax paper, plastic wrap, and aluminum foil all work well. Materials that absorb water, such as paper, probably caused the bubbles to pop because they quickly soaked up the water in the bubble. You might have been surprised to find other seemingly smooth, waterproof surfaces, such as a plastic container or a steel frying pan, also couldn't catch the bubbles without popping them. Remember that even if these materials seem similar to you (for example, an aluminum pan and a steel pan might look very similar and both feel smooth), they might have microscopic differences in their surfaces you can't see—and they are made from different molecules that may be more or less hydrophilic (attracted to water). You might have been even more surprised to find some rough or absorbent surfaces such as carpet could also catch the bubbles. Carpet is composed of lots of individual tiny fibers. When a bubble lands on the carpet it touches many of these tiny fibers at once, so they can hold the bubble up without popping it as a single fiber would.

When you got the surfaces wet, especially with soapy water, it should have become much easier to catch the bubbles. The water forms a thin film on top of the solid surface, preventing the bubble from touching the solid directly. This can greatly extend its life.

CLEANUP

Use a damp paper towel to wipe soap residue off any surfaces.

Stick with It

Put Your Duct Tape to the Test

STICK WITH IT! LEARN ABOUT THE STICKY FORCES BEHIND DUCT TAPE—AND PUT THEM TO THE TEST!

PROJECT TIME

30 to 45 minutes

KEY CONCEPTS

Physics
Adhesion
Van der Waals forces
Wetting

You might have heard of the Apollo 13 lunar mission or seen the movie about the amazing survival of the spacecraft's crew after an explosion crippled the ship on its journey to the moon. What you might not know is that duct tape helped save the astronauts' lives! After the main ship was disabled by an explosion, NASA had to figure out a way to keep the three crew members alive in a tiny lunar lander that was only meant to hold two people. They instructed the astronauts to build lifesaving air filters using cardboard, plastic bags, spacesuit parts and—you guessed it—duct tape! In this activity, we're going to explore how duct tape works—and what can make it stop working.

BACKGROUND

Household glues, such as Elmer's and superglues, are liquid when you apply them. Then, they go through a chemical change as they harden to glue things together. Tape works completely differently—there are no chemical reactions. The stickiness of duct tape (and most other strong household tapes) are in large part the result of two physical properties, known as wetting and van der Waals forces.

Wetting is a word that we use to describe how effectively an adhesive can penetrate (or sink into) the material to which it is applied. Imagine a drop of water on a paper towel and a drop of water on a piece of wax paper. Water dropped on the towel will sink into and penetrate it. Water dropped on the wax paper will form a water droplet on its surface but will not penetrate. Materials that are good at wetting can penetrate into the pores of the materials they're applied to much like the water into the paper towel. The adhesive on your duct tape is good at wetting. When you push a piece of tape down onto a piece of paper, for example, you help the molecules on the sticky side of the tape penetrate into the paper's surface.

Once the adhesive molecules have penetrated the surface of the paper (or other surface), van der Waals forces come into play. These are weak attractions between molecules that normally might not be attracted to one another. Because of the careful arrangement of these molecules in the tape, the tape molecules are able to form physical bonds with the paper molecules. The tape molecules also cohere (stick to one another)—as a result, the tape can hold together and stick to a huge variety of surfaces.

MATERIALS

- Roll of duct tape (at least 5 feet [1.5 m] in length)
- 1/2 cup (118 ml) of water
- Freezer
- Glass dish
- Small baking tray or plate (at least 12 inches [30.5 cm] across)
- About 1 foot (30.5 cm) length of wax paper
- Paper
- Ruler or measuring tape
- Scissors
- Small sandwich bag
- Two large paper clips
- About 15 nickels (You can substitute another coin such as dimes or quarters, as long as you have 15 of them.)
- Timer or clock
- Pen or pencil
- Spoon
- Partner

PREPARATION

- Cut one 12-inch (30.5-cm) piece of tape. Carefully place this piece sticky side up on your tray or plate.
- Put this tray in the freezer, and set your timer for 25 minutes (or note the time on your clock).
- Use a paper clip to carefully poke a hole through a top corner of your sandwich bag. Make your hole at least 1 inch (2.5 cm) from the edge of the bag. Loop your paper clip through the hole (you might have to bend it slightly) so that the bag hangs from the paper clip.
- On your paper, make a table with four columns and five rows. Across the top row write labels for the columns: "duct tape treatment/condition" in the first box; "number of nickels" in the second box; "length of hanging tape" in the third; and "stickiness of tape" in the fourth. In the "duct tape treatment/condition" column write "glass pan" in the row below, "wax paper" in the next row, "wet" in the following one, and "frozen" in the final row. You will fill in the remaining boxes as you do the activity.
- Use your scissors to cut a 12-inch (30.5-cm) strip of tape.
- Gently stick the end of this tape to a countertop or other sturdy object on which you have permission to use duct tape. Allow it to hang free without touching anything. This is your tape control—you will use this to compare the stickiness of the other tape you will use in this activity.

PROCEDURE

- Use your scissors to cut another 12-inch (30.5-cm) strip of tape.
- Stick 6 inches (15 cm) of the tape to the bottom of your glass pan. Press it down firmly using your hand. Allow the other 6 inches (15 cm) to hang below the dish (like a tail!).
- Have your partner hold the dish so that the tape can hang freely.
- Use your scissors or the extra paper clip to poke a small hole through the hanging end of the tape (approximately 1 inch [2.5 cm] from the end) of the tape.
- Insert your paper clip that is attached to the bag through the hole, so that both clip and bag hang from the end of the tape.
- While your partner holds the glass dish steady, slowly add nickels to the bag, pausing to observe any changes after adding each coin. *What happens to the tape as you add nickels? Do you see any changes?*
- If the tape pulls off the glass dish before you run out of nickels, record the number of nickels that the tape held in the first column of the "glass pan" row in your table. If you are able to put all 15 nickels in the bag and the tape holds, record the number 15.
- Use your ruler to measure the tape hanging from the dish. Note the tape's length in the "length of tape" column for the control row in your table. Notice whether the tape's length has changed since you started. *Did the weight of the coins pull more of the tape free, making the tail longer?*
- Gently pull the tape off the glass plate. As you pull the tape, feel the sticky side. Compare this to the hanging half of the tape, which never touched the pan. *Does one half feel more or less sticky than the other? Does it look different?*

- Compare the stickiness of the used tape to that hanging from the table. *Does it feel more or less sticky?*

- Rate the stickiness of the used tape on a scale of 1 to 10. Use your control tape as a comparison. *If that is a 10, how sticky is the used tape?* Dispose of your used tape.

- Use your scissors to cut another 12-inch (30.5-cm) strip of tape.

- Repeat the above steps but this time attach the tape to your wax paper. When your partner holds the wax paper, have them brace it against a book or your glass pan, so that it doesn't sag with the weight of the coins.

- If the tape doesn't come off the wax paper as you pull it, write "could not remove" in the "stickiness of tape" column.

- After you have recorded your results in each column of the wax paper row you can dispose of your used tape.

- Use your scissors to cut another 12-inch (30.5-cm) strip of tape.

- Before attaching the tape, carefully place it sticky side up on a counter or tabletop. Place your ruler next to the tape.

- Use your spoon to drop several spoonfuls of water onto half the length of the tape. Make sure that half of the tape is thoroughly wet.

- Repeat the testing steps, attaching the wet half of the tape to your glass pan. If the tape doesn't stick, write "0" in the "number of nickels" column and "did not stick" in the "length of tape" column.

- Compare the stickiness of the wet half of your tape to the control tape. Rate the stickiness of the wet tape on a scale of 1 to 10 in the "stickiness of tape" column, in the "wet" row. *Does the wet tape feel different? In what way? Does it look different? How would you explain the differences you observe?*

- After you have recorded your results in each column of the wet row, dispose of your used tape.
- Remove the plate with your tape on it from the freezer (after it has been there at least 25 minutes).
- Repeat the testing steps attaching the frozen tape to the glass pan. If the tape doesn't stick, write "0" in the "number of nickels" column and "did not stick" in the "length of tape" column.
- Review your table. *In which condition was the tape able to hold the highest number of nickels? The lowest? Which condition had the highest rating for tape stickiness? Which condition had the lowest?*

EXTRA

Repeat the activity, sticking the tape to a nonstick pan. Be sure to ask for permission first! Add a row to your table to record the results.

SCIENCE FAIR IDEA

Test other types of tape. *Are some more sensitive to water or temperature than others? Do some hold better than others?*

OBSERVATIONS AND RESULTS

During this activity, you should have found the stickiness of your tape depended on two general factors: the condition of the tape and the surface where you were attaching it. Although the tape probably stuck firmly to the glass pan, when the tape was wet or frozen, it was significantly less sticky compared with your control tape. In addition, when you stuck the tape to wax paper, it probably did not stick as well as it did to the glass pan.

So let's start with the conditions of the tape itself: Why doesn't tape work when it's wet? Tape loses its stickiness when the adhesive molecules in the tape become wet. When this happens, the water acts as a barrier to wetting (so the adhesive molecules can't sink into the glass pan) and also prevents the physical bonds from forming between the tape and pan—the water acts like a wall between them, and they can't join together.

Freezing the tape also decreased its adhesive wetting capacity. Again, imagine a few drops of liquid water dropped onto a paper towel, this time compared with an ice cube dropped onto a paper towel. The liquid water will penetrate the towel much more quickly and effectively than the ice cube will. Similarly, the adhesive molecules in the tape, when frozen, cannot penetrate other surfaces as well, so the tape can't get a firm "grip" on the things on which you want it to stick.

In the case of wax paper, the tape didn't stick as well as it did to the glass pan, but this time it was because of the wax paper. Think about the water drop on the paper towel compared with the drop on the wax paper. Just as the water drop can't penetrate the wax paper, neither can the tape! Wax paper blocks most materials from wetting, which makes it great at repelling water, food, and, in this case, tape.

CLEANUP

Throw away any used materials that cannot be reused. You can remove any residue left by the tape using olive oil or an olive oil-soaked paper towel. Allow the oil to saturate the residue for at least 15 minutes, then scrub it off using a sponge.

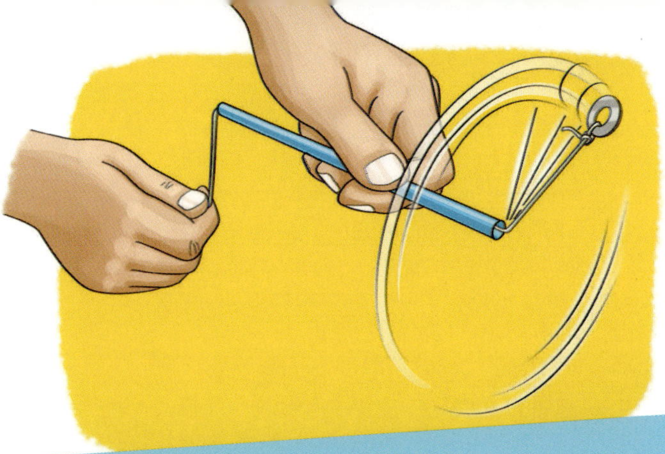

Conserving Angular Momentum
Without the Skates

GAINING MOMENTUM: TEST THE FORCE OF A SPINNING OBJECT WITH THIS FUN PHYSICS ACTIVITY!

You might have seen figure skaters spinning around quickly and then tucking their arms to spin faster, or opening them to spin more slowly. This happens, thanks to a physics concept known as the "conservation of angular momentum." But how can you try it out if you don't have any ice skates around? Find out in this simple activity that only requires common household supplies!

PROJECT TIME

10 minutes

KEY CONCEPTS

Physics
Angular momentum
Tension

BACKGROUND

Have you ever heard of Isaac Newton's first law of motion? It states that an object at rest (not moving) will remain at rest and a moving object will keep moving—unless they are acted on by an outside force. This means that an object's momentum (its mass times its velocity) will stay the same unless an outside force acts on it somehow. You experience this every day. For example, imagine catching a ball. When the ball is moving, it has momentum. You have to exert a force with your hand to stop the ball (and bring its momentum down to zero). If the ball is heavier or moving faster, it has more momentum, and is harder to catch—you have to exert a bigger force.

This concept is usually described in relation to objects moving in a straight line, but it also applies to spinning objects. The terminology is a little different when we talk about spinning objects. Instead of forces, which are pushes or pulls that act in a straight line, we refer to torque, which is a "twist" applied to an object (think: twisting a screwdriver or a doorknob). Instead of mass, we refer to moment of inertia, which measures how spread out the mass is about the point of rotation. Instead of velocity we say, "angular velocity," which measures how fast an object rotates. Finally, instead of momentum, we say "angular momentum."

Even though the terms are a little different, the same concepts apply. The angular momentum of a spinning object will remain the same unless an outside torque acts on it. In physics, when something stays the same, we say it is conserved. That's where the phrase "conservation of angular momentum" comes from. The classic example of this is a spinning ice skater or someone spinning in an office chair. By pulling in her arms, the skater decreases her moment of inertia (all her mass is closer to the middle), so her angular velocity has to increase in order to keep her angular momentum constant.

This can be difficult to demonstrate if you don't have ice skates or a spinning office chair handy, however. In this project, you will demonstrate conservation of angular momentum using a straw, string and small object.

MATERIALS

- Straw
- Washer or other small, heavy object
- Scissors
- String

PREPARATION

- Cut the string to a length of about 40 inches (1 m).

PROCEDURE

- Thread the string through the straw.
- Tie one end of the string to the washer.
- Hold the straw in one hand, and the free end of the string in your other hand.
- Let the washer dangle down from the other end of the straw.
- Gently twirl the straw so the washer spins in a circle. Make sure you spin it fast enough that there is no slack in the string.
- Slowly pull the string through the straw, bringing the washer closer to the straw. *What happens?*
- Now try the opposite. Start with the washer pulled in very close to the straw. Start twirling it, then gradually let the string get looser so the washer can move farther away. *What happens?*

EXTRA

Cut a longer length of string, and go outside into an open space so you can twirl the washer over your head without hitting anything. *Do your results change at all?*

OBSERVATIONS AND RESULTS

When you start twirling the straw with the washer far away from the center and then pull the string through the straw, you should notice that the washer spins faster. This happens because angular momentum is conserved. Angular momentum is an object's moment of inertia times its angular velocity. If one of the latter two increases, the other decreases, and vice versa. When you pull the string through the straw, bringing the washer closer, you decrease its moment of inertia. (That is, you bring its mass closer to the point of rotation.) To compensate and keep angular momentum constant, the washer's angular velocity has to increase. The reverse happens when you loosen the string and allow the washer to move out farther. Its moment of inertia increases, so for angular momentum to remain constant, the rotational speed has to decrease.

CLEANUP

Return all materials to where you got them. Throw away any materials that cannot be reused.

Plug a Leaky Bottle with the Power of Air!

HOW CAN AIR EXCHANGE WORK TO STOP A LEAK? GRAB A BALLOON AND A WATER BOTTLE AND USE PHYSICS TO TRY THIS CHALLENGE!

PROJECT TIME

20 minutes

KEY CONCEPTS

Physics
Air pressure
Volume
Forces

You might have heard people talking about pressure before: maybe that there is too much of it, or not enough. Pressure can mean different things in everyday English, but in general when scientists talk about pressure they mean a constant physical force (such as a push or a pull) exerted or applied to an object. For example, if you sit on an inflated balloon, you can see the effect of the pressure on that balloon—at least, until it pops! In this activity, we're going to use pressure in an unexpected way. Can you keep air in a balloon without tying the bottom? What about keeping water inside a bottle, even if there's a hole in it? Maybe you can do both—with just the right amount of pressure!

BACKGROUND

Have you ever used your finger to seal the top of a straw, and noticed that some of your drink stayed inside the straw—even if you lifted it above the surface of the liquid? If so, then you've already tested the amazing properties of air pressure!

Even though we can't see them, air molecules comprise the air all around us. You can feel them moving if you clap your hands close to your face and feel a small blast of air on your cheek. Those are air molecules! Despite being invisible to us, those molecules still exert a pressure or force on everything in our environment, including you. You can feel their pressure by holding a piece of paper in one hand, facing up toward you. Move the paper up and down quickly. Can you feel the resistance, or force, of the air against the paper? Now try holding the paper with the edge facing you. Move it up and down again, quickly. It was easier the second time because the paper wasn't facing the direction in which you were moving it—there was less resistance or force from the air molecules.

The force of air molecules is important in this activity. They will act on the air and water inside a bottle in a few different (and surprising) ways. Get ready to turn up the pressure!

MATERIALS

- One personal-size plastic water bottle
- A sharp thumbtack or nail
- Scissors
- Two balloons
- Sink
- Measuring tape (A ruler will work too.)
- An adult helper (This activity requires two sets of hands as well as the handling of sharp objects.)

PREPARATION

- Ask your adult helper to use the thumbtack or nail to poke a hole in the side of your water bottle at the very bottom. Have your helper use the scissors to carefully widen the hole, so that it is slightly smaller than the circumference of a yellow pencil.

PROCEDURE

- Hold your finger over the hole in the bottom of the bottle to seal it. Keep the bottle over the sink and fill it with water until there are about 2 inches (5 cm) of empty space at the top. Keep your finger over the hole so no water leaks out!

- Have your helper blow up the balloon until it is approximately 7 inches (18 cm) in diameter. Use the measuring tape or ruler to measure.

- Then have your partner use their fingers to hold the balloon sealed closed, but leave about 1 inch (2.5 cm) of space at the bottom. Keeping it sealed, have them use their other hand to put the mouth of the balloon over the top of the water bottle. **Note:** This can be tricky! Try to do it slowly, easing the balloon over the top of the bottle gently.

- Once the balloon is on the bottle, your helper can let go of it. Measure the diameter of the balloon with your ruler. Keep your finger over the hole! *What happens to the balloon when your helper stops holding it? Does it start to lose air when their fingers are no longer sealing it? Is this surprising? What usually happens when you let go of a balloon without tying it? Why do you think something different is happening now?* (**Note:** If your balloon is shrinking in size at this point, it may have a hole in it. Check for holes before continuing with the activity!)

- Have your helper hold the measuring tape or ruler up behind the balloon at its widest point, so you can measure its diameter. Keep the measuring tape or ruler there for the next step.

- With the bottle still over the sink, lift your finger and allow water to flow out of the bottle. Watch the balloon. When approximately three-quarters of the water has leaked out of the bottle, put your finger over the hole again.

- Have your helper measure the diameter of the balloon. *Has its diameter changed? If so, has it gotten larger or smaller? If not, why do you think the air is staying inside the balloon instead of escaping?*

- Wait 30 seconds and have your helper measure the diameter of the balloon again. *Has the diameter of the balloon changed? If so, has it gotten larger or smaller? If not, why do you think the air is staying inside the balloon instead of escaping?*

- Lift your finger again and allow the water to flow out of the bottle into the sink. *As the bottle empties, does the water seem to flow out of the bottle faster, slower, or at the same rate as it did when you started? What do you notice about the size of the balloon? Do you notice anything else? For example, does the balloon make any noises as the water bottle empties?*

- When all of the water is out of the bottle, measure the diameter of the balloon. *Did the balloon get larger or smaller as the water leaked out of the bottle? Why do you think this happened?*

- Have your helper remove the balloon from the bottle and get ready to try the same test again—this time without inflating the balloon.

- Hold your finger over the hole in the bottom of the empty bottle to seal it. Keep the bottle over the sink and fill it with water until there are about 2 inches (5 cm) of empty space at the top. Keep your finger over the hole so no water leaks out!

- Have your helper put the balloon over the top of the water bottle. *Notice the appearance of the balloon—does it have any air in it at all? How can you tell?*

- With the bottle over the sink, lift your finger and allow water to flow out of the bottle. This time, pay attention to the balloon and the water. When the water level inside the bottle gets down to the halfway point, put your finger over the hole again. *What do you notice about the appearance of the balloon? Has it changed? If so, in what way? What do you notice about the water? Is it flowing out of the bottle as quickly as it did during the first test?*

- Lift your finger again and allow the water to flow out of the bottle. *As it empties, does the water seem to flow out faster, slower, or at the same rate as it did when you started? Does the water keep flowing out of the bottle until it is empty? How would you explain the changes you observe between this test and the one with the inflated balloon? What do you notice about the appearance of the balloon? Does it look different than when you started? In what way?*

EXTRA

Try inflating the balloon to different sizes. Start small and then work your way larger. *Does changing the balloon size affect the results of your test? Why do you think that might be?*

EXTRA

Test taking the deflated balloon off the water bottle at a different point during the second test. *What happens when the balloon isn't on top of the bottle?*

EXTRA

Ask your adult helper to poke a hole closer to the top of the bottle. *How does this change your results?* (Save this one until last!)

OBSERVATIONS AND RESULTS

During the first test, when your finger was covering the hole in the bottom of the water bottle, you should have observed that the balloon stayed the same size and did not leak air. This might have surprised you; usually if you let go of an inflated balloon without tying it, you end up chasing after your balloon as it quickly loses air! What was holding the air inside the balloon this time? The answer is: air (and water)! Most of the room in the bottle was taken up by water, with some air at the top. There might have been a little room at the top of the bottle for more air to squeeze in (so your balloon might have shrunk a small amount in the very beginning), but then it should have stayed the same size. The air in the balloon had nowhere to go, because the bottle was already full!

When you lifted your finger and allowed the water to flow out of the bottle, however, something different happened. You should have seen your balloon start to deflate, or lose air. Why did it lose air when you lifted your finger? Because the water leaving the bottle left room for the air inside the balloon to enter the bottle. Even though you couldn't see it, as the water leaked out, the bottle was slowly filling with air from the balloon. You might have seen this when you covered the hole again. If the balloon was shrinking before you re-covered the hole, it probably stopped when the hole was sealed again. When the water stopped leaving, no more air could fit into the bottle.

In the second part of this activity, you might have noticed that the appearance of the balloon changed as the water leaked out of the bottle. It might have become shriveled looking, as the tiny bit of air inside the balloon went into the bottle. The reason for this result is similar to the explanation for the first test. As the water left the bottle, it created space for the air inside the balloon. Even though in this case it was a small amount of air, it still had an effect on the size of the balloon—and as that air left, the balloon shrunk. In addition, in the second part of the activity you probably noticed a big change in the water leaking out of the bottle. At some point, the water should have stopped leaking out of the hole (or significantly slowed), even though there was still water in the bottle! How is this possible? Again, the answer has to do with air pressure. Because the balloon is empty, no air is trying to get into the bottle from the top. In fact, the balloon seals the bottle, so no air can get in through the top of the bottle. This doesn't mean air isn't exerting a force on the water in the bottle, however. Air molecules outside the bottle are exerting pressure from all sides of the bottle, including at the hole in its bottom.

Normally, without the balloon on top, the water would leak out of the bottle because the pressure from the air coming in the top of the bottle, combined with the force of gravity pulling on the water, is greater than the force of the air molecules pushing on the bottom of the bottle. When we put the balloon on top, however, the air molecules at the top of the bottle can't exert any pressure. The balloon is sealing the bottle. Therefore, the only forces acting on the water are gravity and the force of the air molecules pushing up. In this case, the force of the air outside the hole is great enough to help slow the leakage of water from the bottle.

CLEANUP

Throw away any used materials that cannot be reused. Return materials to where you found them.

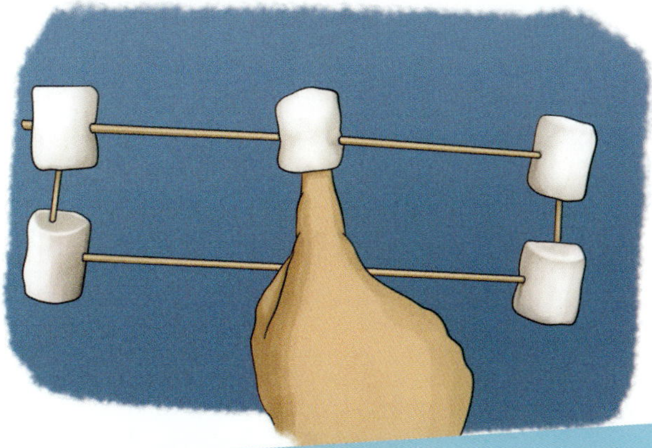

Circus Science

How to Balance Anything

HAVE YOU EVER SEEN A BALANCING ACT? FIND OUT HOW PHYSICS CAN HELP YOU BALANCE MORE THAN JUST CIRCUS ACTS.

Have you ever wondered why it is harder to keep your balance with a heavy backpack on? Or why it is difficult to make a toddler's sippy cup tip over? Or how tightrope walkers manage their impressive circus acts? Or perhaps you want to know how to make your toy car less prone to toppling over when racing through a sharp curve. In this activity, you'll get to investigate balance using marshmallows, skewers, and toothpicks. Read on for some sticky, yummy balancing fun!

PROJECT TIME

20 to 30 minutes

KEY CONCEPTS

Physics
Balance
Mass

BACKGROUND

The balance of an object has everything to do with the distribution of its mass. For example, you might find it easy to stand upright on the edge of a curb when you're holding a heavy backpack down in front of you, but it gets quite a bit more challenging when that backpack is high up on your back. This has to do with how your combined center of mass with the backpack changes in these scenarios.

In physics, the center of mass of an object is a point where the entire mass of the object may be represented as being concentrated. It's the weighted average of the mass particles of the object. It will be in the center of the object if the mass is uniformly distributed (such as with a hula hoop) but shifted to the heavier side of an object if the mass is not uniformly distributed (such as with a car). In some cases, the center of mass is a point in space outside the actual object. For example, the center of mass for a hula hoop is in its empty middle.

MATERIALS

- Toothpicks
- Large marshmallows
- Paper and a pen or pencil
- Two wood or bamboo skewers (use caution when handling skewers with sharp ends)

PREPARATION

- On a skewer, place one marshmallow in the middle and one on each end. Place the tip of your finger under the middle marshmallow and try to balance the structure on your finger. *Can you balance it? Is it easy or difficult?* If you cannot balance it, move the middle marshmallow a bit to one side or the other until you find just the right spot that enables you to balance the structure with your finger. We will call this spot the original balance point.

- Poke a toothpick into each end marshmallow. Add one marshmallow to the end of each toothpick and connect the two new marshmallows with another skewer so you get a rectangular shape (with a marshmallow on each corner and one marshmallow at the center of one skewer).

PROCEDURE

- Place your finger under the original balance point on your new rectangular structure (with most of the structure hanging below this point) and try to balance it on your finger. *Can you balance it? Is it easier or more difficult than balancing the simpler structure?* If you cannot balance this rectangle, move the middle marshmallow a bit to one side or the other until you can.

- Draw the structure on a piece of paper and then try to balance the structure by putting your finger under the original balance point again. *Is the structure completely horizontal, or is it tilted to one side?* Rotate the paper your drawing is on so that it has the same angle as your actual structure. On your drawing, mark where your finger is with an "X" and draw a vertical line down from the X.

- Now balance the structure by putting your finger under one of the toothpicks. *Is the structure hanging down completely vertically or is it tilted?* Rotate the paper your drawing is on so that it has the same angle as your actual structure. Make an X where your finger is and draw a vertical line down from the X.

- Now balance the structure by putting your finger under one of the two corner marshmallows that are farthest away from the middle marshmallow. *Is the structure hanging down completely vertically, or is it tilted?* Rotate the paper your drawing is on so that it has the same angle as your actual structure. Make an X where your finger is and draw a vertical line down from the X.

- On your drawing, the area where the three lines intersect is the center of mass of your structure. *Where is the center of mass of your structure? Where is it in relationship to the middle marshmallow?*

- Reorient your structure so that the middle marshmallow is on top. On the middle marshmallow, use toothpicks to attach two more marshmallows, clustered near the middle one. Place your finger under the original balance point and try to balance it. *Can you balance it? Is it easier or more difficult than balancing the previous structure?*

- Repeat the drawing steps you did before to determine where the three lines intersect and figure out where the center of the mass is now. *Based on your new drawing, where is the center of mass of your new structure? Where is it in relationship to the middle marshmallow? If it has changed, how has it changed? Why do you think this is?*

EXTRA

Try building some other structures using marshmallows, toothpicks, and skewers. *Where is the center of mass for your other structures? Can you alter it by adding marshmallows to specific places?*

EXTRA

Find some non-symmetrical objects, such as a toy truck, and see how their distribution of mass affects how stable they are. *If you park the toy truck on a hill, how will its distribution of mass influence its balance?*

EXTRA

As you did at the beginning of this activity, take a skewer and place one marshmallow in the middle and one on each end. Use toothpicks to stack marshmallows on top of the middle marshmallow and try to balance the structure on your finger each time you add another marshmallow. *How many marshmallows can you add until you can no longer balance the structure?* Make other structures and similarly try stacking marshmallows until they are no longer stable. *Can the other structures hold more stacked marshmallows, indicating that their basic structures may be more stable?*

OBSERVATIONS AND RESULTS

Did you find that the center of mass of the original rectangular structure was just below the middle marshmallow? When you added two more marshmallows to the middle one, did you find that the center of mass shifted up?

An object will more stably balance if a vertical line from its center of mass passes through the balance point. In this activity, the balance point was the spot you marked with an X, where your finger was, and the vertical line is the same as the one you drew. (This vertical line is also known as the plumb line.) By finding where the vertical lines intersected, you could figure out where the structure's center of mass was. You should have found that the first rectangular structure had a center of mass below the middle marshmallow, whereas the second rectangular structure's center of mass was higher up, within the middle marshmallow. Adding two more marshmallows on top of the middle marshmallow should have shifted the center of mass even higher up. Additionally, an object should be more stably balanced if its center of mass is under the balance point, but unstable if the center of mass is above the balance point. Consequently, the first structure you made (with just three marshmallows) should have been less stable than the rectangular structures and, likewise, the original rectangular structure should have been more stable than the one with two marshmallows added.

CLEANUP

Put away all materials. Throw away anything that cannot be reused. Eat the marshmallows if you like!

Shipping Science

Building a Boat that Can Carry Cargo

USE FOIL AND WATER TO SEE HOW A HEAVY LOAD FLOATS WITHOUT FALLING.

Have you ever wondered how a ship made of steel can float? If you drop a steel bolt into a bucket of water, the bolt quickly sinks to the bottom. Then how can a steel ship float? And better yet, how can a steel ship carry a heavy load without sinking? It has to do with the density of the ship (including its cargo) relative to the density of water. In this activity, you'll make little "boats" out of aluminum foil to explore how their size and shape affects the amount of weight they carry and how this relates to water density.

PROJECT TIME

30 minutes

KEY CONCEPTS

Water
Density
Physics
Hydrodynamics
Fluid dynamics

BACKGROUND

What determines whether an object floats or sinks? It's the density (mass per unit of volume) of the object compared with the density of the liquid it is in. If the object is denser than the fluid, the object will sink. If the object is less dense, then it will float.

With a steel-hulled ship, it is the shape of the hull that determines how well it floats and how much of a load it can handle. On an empty ship with a steel hull enclosing a volume of air, the ship's density is equal to the sum of the mass of the steel hull and the mass of the enclosed air, all divided by the hull's volume: The ship floats because its density is less than the density of water. But when cargo or other weight is added to the ship, its density now becomes the sum of the mass of the steel hull, enclosed air, and cargo, all divided by the hull's volume. If too much weight is added, the ship's density becomes greater than that of the water, and it sinks. Excess cargo would need to be thrown overboard in a hurry, or it would be time to abandon ship!

MATERIALS

- Aluminum foil
- Ruler
- Tape
- Calculator
- Scrap piece of paper and pen or pencil
- Bucket, tub, sink, or dishpan
- Water
- Pennies. You may need as many as 200, depending on the size and shape of the boats you make.
- Rag or paper towels
- Dry rice and measuring cup (optional)

PREPARATION

- Cut two squares of aluminum foil. Make one square twice the size of the other square. For example, you could make one square that is 1 x 1 foot (30 x 30 cm) and make the second square 6 x 6 inches (15 x 15 cm).

- Fold the two aluminum foil squares into two different boat hulls. Try to make them the same shape. For example, you could make them both have two pointed ends (like canoes) or you could make them square or rectangular with straight edges that come up on the sides (more like a barge).

- Make finishing touches to the boat hulls. Make sure they don't have any leaks. If needed, use a little tape to make them stronger. Flatten the hull bottoms. On each, try to make sure the rim is the same height going all around the hull edges. *Why do you think this is important?*

- Calculate the volume of each boat hull. If both hulls are square or rectangular, you can measure the length, width, and height of each and multiply these dimensions together to get its total volume. If parts of the hull have an irregular shape, measure the volume piece-wise and then add these volumes together. Use triangles to approximate any areas of the hull that are curved or angled. *What is the volume of each hull?* Write this down.

- Alternatively, you can use dry rice to calculate the volume of each boat hull. To do this, carefully fill each hull with dry rice so that the rice is level with the top of the hull. Being careful not to damage the hull, transfer the dry rice into a measuring cup. *What is the volume of each hull using the rice?* Write this down.

- Fill the bucket, tub, sink, or dishpan with some water.

PROCEDURE

- Take one of the boat hulls and carefully float it in the container of water deep enough to completely submerge the boat.

- Add one penny at a time. To stop the hull from tipping, balance the load as you add pennies (left to right, front to back—or port to starboard, fore to aft, if you're feeling nautical).

- Keep adding pennies until the hull finally sinks.
- Carefully take out the sunken hull and place it and the pennies on a rag or paper towels. Dump any excess water back into the container.
- Count how many pennies the hull could support before sinking (the penny that sank the hull does not count). *How many pennies could it support?* Write this number down.
- Repeat this process with the other hull. Be sure to only add dry pennies. *Why do you think using dry pennies (instead of wet ones) is important?*
- *Could the larger hull support a lot more pennies than the smaller one?*
- Make sure that the volume you calculated for each boat is in cubic centimeters (cm^3). Convert it if necessary and write it down. (Cubic centimeters are the same as milliliters, or ml.)
- Convert the number of pennies each hull could support to grams. To do this, multiply the number of pennies by 2.5 grams (the weight of a single penny). *How many grams could each hull support?* Write this number down for each hull.
- For each hull, divide the number of grams it could support by its volume in cubic centimeters. This roughly gives you the hull's density. *What was the density of each hull right before sinking? How do you think this relates to the density of water?*

EXTRA

In this activity, you compared boat hulls that had two different sizes but you could repeat this activity trying a wider range of shapes and sizes of boat hulls. *Do you get the same results if you use aluminum foil boat hulls that have a different shape? Do you see a pattern in your results?*

EXTRA

You roughly calculated the density of water in this activity, but your calculation could be made even more accurate. Thinking about what factors affected your calculation, develop a way to more accurately determine the density of water, such as by including the weight of the boat hulls, adding something smaller than pennies to measure how much weight the hulls can carry, and more accurately determining the volume of the hulls. *How close can you get your calculations to the actual density of water?*

SCIENCE FAIR IDEA

Repeat this activity, but this time use liquids (or semisolids) other than water, such as cooking oil, liquid detergent, or even snow. (Check with an adult to make sure it is alright if you use household liquids for your activity.) Make sure not to use any dangerous chemicals, such as household cleaning solutions. If you do not want to use much of the liquid, you can dilute it with water or use a small container (just wide enough to fit your boat hull in) and fill it so it is just a little deeper than the height of the hull—so that the hull can sink. *What are the densities of other liquids and semisolids? How do their densities compare with that of water?*

OBSERVATIONS AND RESULTS

Did the larger boat hull support a greater amount of weight? Did both hulls have a similar density right before sinking, which was roughly 1 gram per cubic centimeter—the density of water?

When you first put one of the boat hulls on the water, it should float because its total density is less than that of water. As you add pennies to the hull, its density increases and the hull floats lower. Eventually, when enough pennies are added, the hull's density roughly equals the water's. This happens right before the penny that sinks the hull is added. The hull sinks because its density has finally become greater than that of water. Consequently, the hull's density right before sinking should roughly equal the density of water, which is 1 gram per cubic centimeter. Even though the larger hull supports more weight, it also has a greater volume, and both hulls should roughly have a density of 1 gram per cubic centimeter right before sinking. (Your densities may not have been exactly this, but may have ranged between 0.7 to 1.3 grams per cubic centimeter. Sources of error that could be eliminated to give you an answer closer to the actual density of water include more accurately calculating the volume of each hull, using something smaller than pennies and including the hull's weight in your calculations.)

CLEANUP

Be sure to recycle the aluminum foil when you are done testing your hulls.

Singing Glasses

GLASSES AS INSTRUMENTS? FIND OUT HOW GLASS AND WATER MAKE BEAUTIFUL MUSIC!

PROJECT TIME

20 to 30 minutes

KEY CONCEPTS

Sound
Physics
Music
Resonance

Thanksgiving can be a wonderful time of year, with friends and family getting together. The table is full of delicious food—a big feast—and there might even be some fancy dishes and glassware on the table to make the celebration even more special. But did you know that you can make wine glasses sing? Perhaps you have done this many times before, but have you ever wondered why the wine glass makes this sound? Do this activity, and you will not only make wine glasses sing, but you will also learn how to generate different notes; you might even get inspired to play a little song.

BACKGROUND

Glasses have been used for making music since the Middle Ages. The first musical instrument made of upright wine glasses was called the glass harp and was invented around 1750. About 10 years later, Benjamin Franklin—one of the Founding Fathers of the United States—invented a mechanical version of the glass harp, called the glass harmonica. Both of these instruments are based on the principle of generating musical tones by means of friction. What does that mean?

When you rub your moistened finger along the rim of the glass, your finger will stick to the glass as it encounters resistance, or friction, when it moves over the glass surface. The water on your finger, however, will allow your finger to slip, as it forms a cushion that reduces friction. When the pressure and amount of moisture are just right, this so-called stick-slip motion (the slight friction between your finger and the rim of the glass) will cause vibrations in the sides of the glass. The sides of the glass transmit the vibration to the surrounding air, creating a sound wave with a specific frequency. The frequency specifies the rate at which a vibration occurs and is usually measured per second or Hertz (Hz). There is a particular frequency, called the resonant frequency, at which the sides of the glass will vibrate most easily. The resonant frequency of wine glasses is typically within the range of human hearing (20 to 20,000 Hz), and this is why you hear the resulting resonant vibration as a tone. Now let's make some music, and you can play your very own wine glass instrument.

MATERIALS

- Wine glasses and permission to use them carefully (optional: different size and shapes)
- Water
- Graduated liquid measuring cup
- Metal spoon
- Ping-Pong ball (optional)
- Chromatic tuner (optional, available as free phone app)
- Piano or keyboard for comparing notes (optional)

PREPARATION

- Find a working area that can tolerate some water spills.

- Carefully gather your wine glass(es) and your measuring cup, as well as some water.
- Wash your hands with dish soap and rinse your hands thoroughly.

PROCEDURE

- Take an empty wine glass and put it in front of you on the table.
- Hold down the base of the glass with your non-dominant hand.
- Rub your dry finger around the rim of the glass while pressing down gently. Observe how it feels. *Does your finger stick to the glass or is it gliding smoothly along the rim? Do you hear a sound?*
- Wet the index finger of your dominant hand with water.
- Rub your finger around the rim of the glass again while pressing down gently. *Does it feel different than with a dry finger? How much resistance do you feel this time? Does the glass start to sing?* **Tip:** If you hear no sound, try changing the speed and applying more or less pressure to the rim.
- While you hear the glass sing, stop rubbing your finger along the rim and take it away from the glass. *Does the glass continue to sing or does it stop?*
- Now make the glass sing again and stop rubbing your finger along the rim but this time leave your finger on the rim on the glass. *What happens to the sound this time?*
- Take a metal spoon and very gently tap the glass on the side. *Does this generate a different sound or note compared to when you used your finger to make the glass sing?*

- If you are using a chromatic tuner, use it to determine the note the glass makes when rubbing it with your finger or hitting it with the metal spoon. You can also compare the note frequency with notes sounded on a piano or keyboard. *What note did the glass generate? Can you sing the same note?*

- Now fill one-third of the glass with water and repeat the procedure. *What note does the glass make now? Is it a higher or lower pitch?*

- Watch the water surface in the glass carefully while you are generating the sound. *Does the water surface stay calm or do you see any disturbances?*

- Take the metal spoon again and very gently tap the glass on the side. *Do you get the same note again or is it different?*

- Again, you can use the reading from the chromatic tuner to determine the note the glass makes when rubbing it with your finger or hitting it with the metal spoon. You can also compare the note frequency with notes sounded on a piano or keyboard. *What note did the glass generate this time? Is it different from the previous one? How?*

- Add more water to the glass so that two-thirds of the glass is filled with water. Rub the rim of the glass again with your damp finger. *How does the pitch of the sound change with the different water level? Did the sound become higher or lower?*

- Take the metal spoon for the last time and very gently tap the glass on the side. *What note do you generate?*

- Compare the three notes that you generated with each water level (empty, one-third full, and two-thirds full). *Can you find a relationship between the water level in the glass and the resulting pitch of the note?*

- If you are using a chromatic tuner, use it again to determine the note the glass makes when rubbing it with your finger or hitting it with the metal spoon. You can also compare the note frequency with notes sounded on a piano or keyboard.

SCIENCE FAIR IDEA

Try the activity with wine glasses of different sizes and/or shapes. *How does the size or shape of the glass change how it sounds? Can you draw any conclusions about the relationship between the size/shape of the glass and its resonance frequency?*

EXTRA

Put one glass close to another wine glass and let it sing. Now hold it so the sound stops. *Does the second glass sing instead?*

SCIENCE FAIR IDEA

Take an empty wine glass and put a lightweight object in it, such as a Ping-Pong ball. Rub your damp finger around the rim of the glass, and once the glass starts making a sound, tilt the glass on its side. Be careful not to drop the Ping-Pong ball on the floor. Continue rubbing the glass to make it sing and observe the ball inside. *What happens to the ball? Does it sit still on the walls of the glass or does it move?*

OBSERVATIONS AND RESULTS

Could you make the wine glass sing with your dry finger? Probably not, as there was too much friction between your finger and the glass rim. To create a stick-slip motion, the finger has to be a little wet. But once the conditions were right, the glass should have generated a clear and beautiful sound. By rubbing along the glass rim with your finger, you made the walls of the glass vibrate; you might even have felt the vibrations in your finger. When you take your finger away from the glass, the glass walls continue to vibrate, thus, still producing a musical note. However, if you stop rubbing the rim and leave your finger touching the glass, your finger will stop the vibrations and the sound stops as well. In a wine glass filled with water, you actually should have been able to see the generated vibrations in the form of a little wave pattern that develops, especially around the glass walls. If you did the extra activity with the Ping-Pong ball, you also probably made the vibrations visible. The ball should dance inside the glass as it gets pushed around by the vibrations of the walls each time it comes in contact with the glass edges.

In each case when you tapped the glass with the metal spoon, you should have heard the same note as when you used your finger. Using the metal spoon is just another method to get the glass walls to vibrate, and the generated sound waves of the glass thus stay the same. Sometimes, even vibrations of one glass can make another glass vibrate and sing if you put them very close together. However, when adding different amounts of water to the glass, you should have noticed that the pitch of the sound changed depending on the water level; the more water you add to the glass, the lower the pitch of the sound you hear. This is due to the fact that the water volume inside the glass makes it much heavier and therefore, it is more difficult for the glass walls to vibrate. That means that the sound wave generated by the vibrations is much slower or has a lower frequency. As the frequency of a note is correlated to its pitch, the pitch produced by the glass gets lower as you add more water. If you can use more than one glass, you can play around with different water levels or glass shapes and sizes; each one will have a different resonance frequency and will produce a different note. Try to find different glasses and change water levels to generate each note of the musical scale. Can you play a simple song with your glass harp?

CLEANUP

Wash and dry any glasses you used. Put them away.

Simple Machine to Make Potato Holes

PENCIL OR STRAW: WHICH CUTS BETTER? TRY YOUR HAND—AND SOME SIMPLE MACHINES—AT THIS SIMPLE TEST, AND THE ANSWER MIGHT JUST SURPRISE YOU!

Looking for a cool and surprising project to show to your friends? Give this activity a stab! Take a potato and a straw—maybe a pencil as well—and challenge friends to drill holes into the potato with the least amount of effort. Chances are, they will not find a method that gets the job done easily. The best method—one that requires the least effort from you—will surprise everyone! How? Investigate and have fun! Can you find a reliable way to get the job done? Can you explain the physics behind it?

PROJECT TIME

20 to 30 minutes

KEY CONCEPTS

Physics
Simple machines
Wedges
Air pressure

BACKGROUND

The human body is equipped with an amazing set of tools: fingers, teeth, arms, and more. Sometimes, however, these are not enough to do the job. We use tools to enhance the human ability to push or pull to make tasks easier. Think about wheelbarrows, scissors, knives, catapults, and screws, all of which help us complete various tasks. Scientists call these tools simple machines. A wedge is one example. It has two planes: One end is thick (such as the back of an ax or top of a needle); it gradually narrows into a sharp edge on the other end (the blade of an axe or a needle point). Wedges are used to separate two objects, split one object, keep objects together or prevent them from moving. Examples of wedges are nails, knives, staples, and door stoppers. Wedges concentrate all the pushing applied on the thick end of the wedge to a smaller surface on the sides or the sharp end or tip. The result is clear—less pushing is needed to get the job done. Try it with a butter knife. Push the cutting-blade end into a brick of cheese, then turn the knife around and carefully try to push the handle tip into the cheese. Which side needs less effort to cut the cheese? The sharper side, right?

A wedge is one of six types of simple machines. The others are pulleys, levers, wheels, inclined planes, and screws. These machines seem quite intuitive, but sometimes surprises happen. See where you can find some wedges in this activity, but don't forget to be creative and get thinking about other factors at play!

MATERIALS

- Potato, at least 1-inch (2.5-cm) thick (Preferably it should be fresh and hard.)
- Cutting board
- Knife to cut the potato (An adult should help with this step.)
- Drinking straws, one for each method you try (Any bendable parts should be carefully removed.)
- Pencil with one pointy side and one flat side (A chopstick with one pointy side works as well.)

PREPARATION

- Ask an adult to cut the potato so you have at least one slice that is about 1 inch (2.5 cm) thick.

- Place the slice(s) on the cutting board.
- Your goal in this activity is to drill holes about the size of a pencil or drinking straw into the potato slice. *What methods can you think of with the materials listed for this activity?* In addition to your own ideas, you can find eight methods to try below. Try each one of them; see which one(s) are easy and try to explain why.
- Here are a few tips that can help you approach this activity like a scientist:

    • Always form a hypothesis (or educated guess) before you try a method. Completing the following sentence will help you create one: "If I ... (describe the method you are about to try), I ... (expect to/do not expect to) create a hole through the potato slice because ...."

    • Use a new straw for each test. Straws might get damaged during a test so to try the different methods under similar conditions, you will need a new straw for each test.

    • After each test, take a moment to investigate the results. *Was your hypothesis correct? Why do you think this is the case? Did you learn anything from this test?*

- For safety reasons, always have the potato on the cutting board while testing.

PROCEDURE

- "Twisting a straw": hold the straw at about three-fourths of the way up, and try to twist the straw in. If you are wondering if the edges of a straw are sharp, rub your finger along an edge. *Does it feel sharp? Do you think it will cut into the potato?* Make a hypothesis, perform the test and investigate the results. *Did it work? Was it easy? What have you learned from this test?*

- "Twisting the flat side of a pencil": hold the pencil about three-fourths of the way up from the flat end and try to twist the flat end into the potato. *Do you think this will work?* Make a hypothesis, perform the test, and see what you learn. *What happened on this test?*

- "Twisting the pointy side of a pencil": turn your pencil around and, holding it three-fourths of the way up, try to twist the pointy side of your pencil into the potato. *Will this be easier or more difficult than previous methods?* Make a hypothesis, perform the test, and investigate the results.

- *Did any of the twisting methods work?* Let us try something different: jamming the item into the potato.

- "Jamming a straw": hold your straw about three-fourths of the way from one end (without pinching the straw) and try to jam the far end into the slice. *Can you force the straw through the potato?* Make a hypothesis, perform the test, and investigate the results. *What have you learned from this test? Was your straw sturdy enough or did it bend under the force?*

- "Jamming the flat side of a pencil": repeat the previous method with the flat side of the pencil.

- "Jamming the pointy side of a pencil": repeat with the pointy side of the pencil.

- "Jamming a plugged straw": back to the straw with the sharp edges. (Use a new straw for this test.) Hold it so your thumb blocks off one end of the straw. Carefully try to jam the other end into the slice. Keep making a hypothesis before you test, and investigate after the test. *Was your prediction correct? Why do you think this works or does not work?*

- "Jamming a folded straw": now, try to hold the straw in yet another way. Fold one end of the straw over and hold it firmly closed so little to no air can escape through this end of the straw. Try to jam the other end into the slice. *Do you think this will work? Did it work? Why did it work or not work?*

- Look back at all the methods you tried. *Which worked well; which worked, but with difficulty; and which did not work at all? Can you see a trend? Can you explain why some methods work extremely well, others are hard, and still others do not work at all?*

SCIENCE FAIR IDEA

Try these methods on other fruits or vegetables or on other materials of varying hardness (with permission, of course, and have an adult supervise you). *Will it work? Why or why not?*

EXTRA

Can you come up with other ways to cut a hole this size through a potato? See if you can use other materials you find around the house (but be sure to ask permission, and have an adult supervise you).

EXTRA

As explained in the background section, the methods that worked for this activity used—among other things—a simple machine called a "wedge" to cut the hole in the potato. *Can you find other examples of wedges around your house?*

OBSERVATIONS AND RESULTS

Jamming a plugged straw worked with amazing ease, right? Jamming the pointy side of a pencil was probably successful, although it needed some more effort; twisting the pencil was likely hard or unsuccessful.

Both the edge of a straw and the pointy side of the pencil are wedges. Wedges make it easier to push aside the potato flesh. On top of that, you used the power of jamming. Whereas twisting provides gradual force spread out over time, jamming provides instant force. The wedge concentrated this instant force on a small surface area—the border of the straw or the pencil point. This allowed the straw or pencil to penetrate the potato.

The plugged straw would most impress your friends if you challenge them with this activity; it is strong enough to penetrate whereas the unplugged version bent. Do you know why? When plugging the straw and jamming the other end on the potato, you trap air inside the straw. This air keeps the straw stiff. It pushes on the walls of the straw, which prevents the straw from folding. Air is a gas, and gases can easily be compressed. It still allows the potato flesh to enter the straw but at the same time prevents the straw from collapsing and bending.

The plugged straw worked better than the pencil because with the straw, you only displaced a very small amount of potato to make room for the straw. With the pencil, although you displaced much more potato, displacing material is work, so it felt more difficult to do.

CLEANUP

The potato can go in the compost bin. Throw straws in the trash, and wash other kitchen equipment with warm water and soap.

Science with a Smartphone Accelerometer

MEASURE YOUR MOVES WITH A SMARTPHONE—
AND A LITTLE PHYSICS.

Have you ever played a video game with a controller that used motion controls? Do you ever wonder how sometimes a smartphone seems to "know" if you're moving? How do these electronic devices measure motion? Try this activity to find out!

PROJECT TIME

30 to 40 minutes

KEY CONCEPTS

Physics
Motion
Position
Velocity
Acceleration
Measurement

BACKGROUND

You're probably familiar with the units we use to measure distance and velocity. In the United States, we might say someone is 5 ft 11 in tall (measuring distance) or that we drive 55 miles per hour on the highway (measuring velocity, which is a unit of distance during an amount of time). Scientists use the metric system, which measures distance in meters (m) and velocity in meters per second (m/s).

To reach a certain velocity, an object will often need to speed up or slow down. You might use the word "accelerate" to describe something that is speeding up (such as a car accelerating after a red light) or "decelerate" to describe something that is slowing down (such as a car applying the brakes). We can explore these ideas even further. Whereas velocity measures how fast distance changes, acceleration measures how fast velocity changes. That makes the units for acceleration sound a little weird: meters per second per second—or meters per second squared (m/s^2). If something has a constant velocity, such as a car using cruise control on the highway, its acceleration is zero. If the velocity changes at all, the acceleration is not zero. For example, if a car is stopped at a stoplight, it has a velocity of 0 m/s. If it accelerates, and two seconds later it has a velocity of 10 m/s, its acceleration was 5 m/s^2. Roller coaster riders can experience accelerations up to approximately 60 m/s^2, and fighter jet pilots might experience accelerations up to 90 m/s^2 for brief periods of time! Don't worry if this all sounds confusing—you can just think of acceleration as how fast something is speeding up or slowing down (if it's slowing down, the number will be negative).

What does all this have to do with smartphones and video game controllers? Acceleration can be measured with a small electronic device called an accelerometer. Most smartphones these days contain built-in accelerometers and can run apps that display the accelerometer readings. So if you want to explore motion in the world around you, all you need is a smartphone!

MATERIALS

- Smartphone or tablet with internet access and permission to download and install an app
- Soft surface, such as a pillow, bed or couch
- Adult (to help verify and download the app)
- Access to a playground (optional)
- Velcro band (or other method) for attaching the phone to your arm or leg (optional)
- Vehicle to ride on or in, such as a bicycle, car, bus, and so on (optional)

PREPARATION

- Ask an adult to help you search for and download an "accelerometer" app on a smartphone or tablet. There are plenty of free options available, but some apps may have ads or in-app purchases enabled.
- Get to know your accelerometer app. Some apps will just display a number on the screen, whereas others will display a meter or a graph. An app that lets you record data and automatically finds minimum and maximum values will work best for this project.
- Most accelerometer apps display three acceleration readings, with labels such as "X," "Y," and "Z." These correspond to the motion of your phone in three-dimensional space.
- Usually if you put your phone flat on a table with the screen facing up, "X" acceleration would be any movement left to right, "Y" acceleration would be any movement in the direction of the top or bottom of the device, and "Z" acceleration would be any movement up and down (above or below the table surface).
- Some apps might give you the option to combine readings from all three directions and will display the combined or total acceleration.
- Make sure the app is working: wave your phone around, and you should see the numbers change.

PROCEDURE

- Measure the acceleration of "typical" motions you make as you go about your day. Hold the phone in your hand or put it in your pocket as you walk around, sit down/stand up, go up and down stairs, and so on. *How big are the accelerations you measure?*

- Now measure the accelerations of "fast" motions. Try jumping, running, twirling, or waving your arms around (be careful not to drop the phone!). *How large are these accelerations?*
- Try dropping the phone a short distance onto a soft surface, such as a pillow, bed, or couch. *What is the acceleration when the phone hits the ground? Do you think the acceleration would be bigger if the phone fell from a higher place or landed on a hard surface?* (Don't test that one!)
- *How difficult is it to move with a constant velocity?* Put the phone down on a table and try pushing it in the X or Y direction. *Can you get the corresponding acceleration reading to stay at zero?*

EXTRA

Try tilting the phone without moving it. *Do the acceleration readings change depending on which way you tilt the phone?*

EXTRA

Take acceleration readings on a playground. *What happens when you go down a slide, swing on a swing, or climb on the monkey bars? Where do you experience the greatest acceleration?*

EXTRA

Take acceleration readings from different parts of your body. You can use a Velcro (or other) band to attach the phone to your arm or leg. *Which part of your body experiences the biggest acceleration when you run? The smallest?*

EXTRA

Take measurements in a vehicle. What are the accelerations when you ride your bike (don't look at the phone while you're riding!) or ride in a car or on a bus?

OBSERVATIONS AND RESULTS

You should easily measure accelerations up to approximately 10 m/s^2 as you go through regular motions while holding a phone. Smartphones, fitness trackers, and pedometers can track changes in these accelerations to count how many steps a person takes throughout the day.

The accelerations might be a little higher if you put the phone in your pocket instead of holding it because it bounces around more. Faster motions, such as swinging the phone, could result in accelerations of more than 50 m/s^2—close to the same accelerations felt by someone riding a roller coaster. Some electronic devices, such as laptops, actually have a "drop detector" and will automatically power down to help prevent damage if they detect accelerations that are too high. (Again, don't test this one!)

You might be confused if you discovered that you could get the acceleration reading to change just by tilting your phone. After all, the phone isn't moving around—its velocity is zero, so shouldn't the acceleration also be zero? This occurs because accelerometers also detect acceleration due to gravity, which is 9.8 m/s^2. As you tilt your phone the direction of gravity relative to the phone's body changes. Even if the phone is holding still, this will make the X, Y, and Z acceleration readings different depending on which one is pointing downward. This is how motion controls in video games (for example, tilting the controller to steer in a racing game) work.

CLEANUP

If you used any additional materials, return them to the place where you found them.

THE SCIENTIFIC METHOD

The scientific method helps scientists—and students—gather facts to prove whether an idea is true. Using this method, scientists come up with ideas and then test those ideas by observing facts and drawing conclusions. You can use the scientific method to develop and test your own ideas!

Question: What do you want to learn? What problem needs to be solved? Be as specific as possible.

Research: Learn more about your topic and refine your question.

Hypothesis: Form an educated guess about what you think will answer your question. This allows you to make a prediction you can test.

Experiment: Create a test to learn if your hypothesis is correct. Limit the number of variables, or elements of the experiment that could change.

Analysis: Record your observations about the progress and results of your experiment. Then analyze your data to understand what it means.

Conclusion: Review all your data. Did the results of the experiment match the prediction? If so, your hypothesis was correct. If not, your hypothesis may need to be changed.

GLOSSARY

density: A measurement of how tightly something is packed together.

flow: The motion of a fluid.

force: A push or pull on an object that changes its speed or direction.

frequency: The number of waves that pass a point during a set amount of time.

friction: A force that resists sliding between two surfaces.

gas: The state of matter made up of particles that are spread far apart.

inertia: The tendency of something to resist change in motion or rest.

liquid: The state of matter that can flow and change shape based on its container.

mass: The amount of substance that makes up an object.

momentum: The power an object has when it is moving.

particle: The smallest possible unit of matter.

pressure: The pressing or pushing over a specific area.

solid: The state of matter that keeps its own shape, no matter the container.

torque: The measure of the force that makes an object rotate on an axis.

velocity: The speed in a particular direction.

volume: The amount of space that an object takes up.

ADDITIONAL RESOURCES

Books

Barfield, Mike. *Feel the Force: Revealing the Physics Secrets that Rule the Universe.* London, UK: Laurence King Publishing, 2023.

VanCleave, Janice. *Janice VanCleave's Physics for Every Kid: Easy Activities That Make Science Learning Fun.* San Francisco, CA: Jossey-Bass, a Wiley Brand, 2021.

Weakland, Mark, and Suzanne Slade, *Wile E. Coyote's Physical Science for Super Geniuses in Training.* North Mankato, MN: Capstone Press, 2023.

Websites

Discovery Education
sciencefaircentral.com

Exploratorium
www.exploratorium.edu/explore/activities

Science Buddies
www.sciencebuddies.org/science-fair-projects/project-ideas/list

Science Fun for Everyone!
www.sciencefun.org/?s=science+fair

Videos

What Can You Do with a Physics Degree?
https://wqed.pbslearningmedia.org/resource/physics-degrees-careers-physicsgirl-1028/what-can-you-do-with-a-physics-degree-physics-girl/

What Is Physics?
https://wqed.pbslearningmedia.org/resource/2420fafa-9d03-4b92-965f-2b014a533868/unit-1-segment-a-what-is-physics-physics-in-motion/

Publisher's note to educators and parents: Our editors have carefully reviewed these websites to ensure that they are suitable for students. Many websites change frequently, however, and we cannot guarantee that a site's future contents will continue to meet our high standards of quality and educational value. Be advised that students should be closely supervised whenever they access the internet.

INDEX